COMMON EDIBLE SEAWEEDS
IN THE GULF OF ALASKA

DOLLY GARZA
Alaska Sea Grant
Marine Advisory Program

Published by the Alaska Sea Grant College Program
SG-ED-46

Elmer E. Rasmuson Library Cataloging in Publication Data

Garza, Dolly A.

    Common edible seaweeds in the Gulf of Alaska / Dolly Garza. — Fairbanks, Alaska : Alaska Sea Grant College Program, University of Alaska Fairbanks 2005.

    p. : ill. ; cm. - (Alaska Sea Grant College Program, University of Alaska Fairbanks) ; SG-ED-46)

    NOAA National Sea Grant NA16RG2321, A/152-20, A/161-01.

    1. Marine algae as food—Alaska—Alaska, Gulf of. 2. Cookery (Marine algae) I. Title. II. Series: Alaska Sea Grant College Program, University of Alaska Fairbanks ; SG-ED-46.

TX402.G37 2005

ISBN 1-56612-086-1

Credits

Cover and book design, and layout, by Dave Partee. Illustrations by Ernani G. Menez, except pages 1, 24, and 39. Cover photos by Stephen Jewett (background), Dolly Garza (top and middle), and Kurt Byers (bottom). Thanks to the Central Council: Tlingit and Haida Indian Tribes of Alaska for a financial contribution. Mike Stekoll, Brenda Konar, and Nora Foster reviewed the text, and Sue Keller edited the book.

    Published by the Alaska Sea Grant College Program, supported by the U.S. Department of Commerce, NOAA National Sea Grant Office, grant NA16RG2321, projects A/152-20 and A/161-01; and by the University of Alaska Fairbanks with state funds. The University of Alaska is an affirmative action/equal opportunity employer and educational institution.

    Sea Grant is a unique partnership with public and private sectors combining research, education, and technology transfer for public service. This national network of universities meets changing environmental and economic needs of people in our coastal, ocean, and Great Lakes regions.

Alaska Sea Grant College Program
University of Alaska Fairbanks
P.O. Box 755040
Fairbanks, Alaska 99775-5040

Toll free (888) 789-0090
(907) 474-6707 • fax (907) 474-6285
alaskaseagrant.org

Contents

The Author iv
About This Book iv
Introduction to Seaweeds 1
Traditional Uses 3
Nutrition 4
Collecting Seaweed 4
    Planning *5*
    Harvesting *6*
    Processing *7*
Alaria marginata (Winged kelp, wakame) 8
Fucus gardneri (Popweed, rockweed) 12
Laminaria (Kelp, kombu) 16
Nereocystis luetkeana (Bull kelp, bullwhip kelp) 20
Porphyra (Black seaweed, nori, laver) 24
Palmaria mollis (Ribbon seaweed, dulse) 30
Ulva fenestrata (Sea lettuce) 34
Salicornia virginica (Beach asparagus) 38
Recipes 41
    Seasonings *41*
    Snacks *42*
    Main Dishes *43*
    Side Dishes *48*
    Canned Products *50*
References 54
Index 56

The Author

Author Dolly Garza on a seaweed-photographing expedition near Sitka, Alaska.

As a Sea Grant Marine Advisory agent, Dolly Garza has taught seaweed identification and use in communities around Southeast Alaska for about 10 years. She grew up collecting and using seaweeds with her family in Ketchikan and Craig, and enjoys harvesting and creating new seaweed dishes. Garza is professor of fisheries at the University of Alaska Fairbanks, and is a Marine Advisory agent based in Ketchikan. She earned her Ph.D. in marine policy from the University of Delaware.

She can be reached in Ketchikan at:

Marine Advisory Office
2030 Sea Level Dr. #201b
Ketchikan, Alaska 99901

Phone: (907) 247-4978
Fax: (907) 247-4976
ffdag@uaf.edu

About This Book

While there are many seaweeds in Alaska's waters, it is the intent of this booklet to share with you helpful information about only a handful of common, abundant seaweeds (and one beach plant) that you may enjoy identifying, collecting, and eating. For more in-depth information on seaweed biology, and for longer lists of species, please see the reference section.

This book would be better titled "Common Edible Seaweeds in the Gulf of Alaska *and One Flowering Plant*." As a bonus, the easily collected, and nutritious, beach asparagus is included for your eating pleasure.

Introduction to Seaweeds

Seaweeds are macroalgae in all oceans of the world. They are found in the nearshore subtidal and intertidal areas along varied coastlines from tropics to arctic areas. Seaweeds are important to nearshore ecosystems because they provide refuge for many invertebrates and fishes, and contribute important organics to the ecosystem.

Thousands of edible seaweeds of various shapes and sizes occupy a wide array of ecological niches. The distribution of seaweeds worldwide and within regions depends on sea temperature, light availability, suitable attachment surfaces, nutrients, and wave action.

You will often see visible bands or zones of seaweeds from high tide down through subtidal zones. A few seaweeds are found in several zones but are more abundant in one zone.

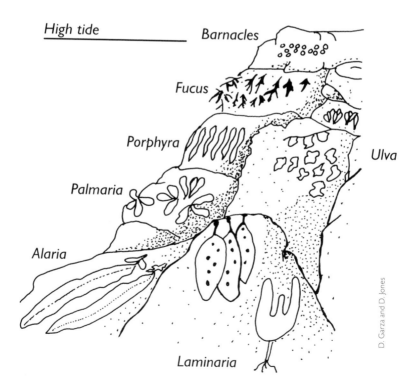

Seaweeds are not considered true plants. While they photosynthesize like plants, seaweeds lack structures like roots, stems, and leaves that provide nutrients and water in true plants.

A seaweed may have a holdfast that attaches the algae to a substrate like a rock, a pliant stipe that looks like a plant stem, a frond or bladelet (or sporophyll) that looks like a leaf, and possibly bulbs or gas-filled sacs which may help to keep the algae floating in the water column, or on the surface.

While seaweeds have complex reproductive strategies that are not immediately observable, you will notice that some seaweeds are found year-round, while others die back and reappear the next spring or early summer.

The three divisions of macroalgae are the Phaeophyta or brown algae, the Rhodophyta or red algae, and the Chlorophyta or green algae.

Phaeophyta: There are approximately 1,500 marine species of brown seaweeds (kelps) worldwide. Most of the brown seaweeds have complex structure with stipes, blades, and visible holdfasts. Some can renew blade tissue up to five times per year. They are found throughout the tidal zone, from *Fucus* in the upper levels to *Nereocystis* in much lower subtidal areas. The browns may contain alginic acid, iodine, and potassium. *Nereocystis* and *Alaria,* among others, form large beds or canopies.

Herring roe on giant kelp is a delicacy for Pacific Northwest Native tribes, including Alaska.

Many brown seaweeds in the Gulf of Alaska that are important to the ecosystem are also good food sources. Common brown seaweeds include *Laminaria*, *Alaria*, *Fucus*, and *Nereocystis*.

Rhodophyta: There are approximately 4,000 species of marine red algae worldwide. They can be found abundantly during a "minus" tide. Red seaweeds can have carrageenan, agar, bromine, and calcium. Two important red macroalgae in the Gulf of Alaska are *Porphyra* (black seaweed) and *Palmaria* (ribbon seaweed).

Chlorophyta: There are fewer than 1,000 marine species of green seaweeds worldwide. The greens are more abundant in warmer waters. One important food species in the Gulf of Alaska is *Ulva* (sea lettuce).

Traditional Uses

Several seaweeds are important to the Haida, Tlingit, Tsimshian, Eyak, and Alutiiq people. *Porphyra* (black seaweed), *Palmaria* (ribbon seaweed), *Nereocystis* (bull kelp), and *Macrocystis* (giant kelp) continue to be used by Northwest tribes. Black seaweed and ribbon seaweed are important food and trade items. In the Bristol Bay region, *Fucus* laden with herring eggs is a treasured spring food. *Macrocystis* covered with herring eggs is a delicacy, and also an important trade item. While herring roe on *Macrocystis* is an important subsistence food, *Macrocystis* is limited in distribution and is not included in this book.

Bull kelp was important in pre-European technology, used by the West Coast Native peoples. Rope was made from the long slender portion of the stipe, which is found near the holdfast. This rope was used only in marine situations such as to anchor something out. In addition the hollow bulb portion was used to store foods in, such as oil from the eulachon, a small herring-like fish.

Nutrition

Seaweeds have various minerals, vitamins, carbohydrates, and sometimes protein. They are very low in fat and are approximately 80-90% water.

There is an abundance of information on the nutrition of seaweeds in books listed in the reference section. While most of the nutrition studies were conducted on non-Alaska samples, the information is still useful. Different seaweed species have different vitamin and mineral content, but most are nutritious.

Collecting Seaweed

Seaweeds usually are picked in the spring and into the summer when abundant light and an influx of nutrients provides for rapid new growth. In spring the early morning light coincides with good low tides, making picking easier. There are exceptions, as some of the larger seaweeds like *Nereocystis* and *Laminaria* can be picked into the fall and even into the winter.

For a seaweed collecting trip you will need wading boots, a light raincoat, warm clothes, and survival gear.

Planning

For a collecting expedition you will need wading boots, a light raincoat, warm clothes, and survival gear. The harvesting gear is simple: a small pair of scissors or a small paring knife (for bull kelp a large knife is needed) and several bags in which to place seaweed. The bags may be old pillowcases, mesh bags, or once-used grocery bags. Stay away from garbage bags as some are treated with chemicals and may contaminate your harvest. A backpack is ideal to hold your supplies and harvest, allowing your hands to be free for climbing around the rocks.

Store your emergency gear including signals, drinking water, etc., in your backpack so you will have easy access to them if you become stranded on a rock. You also want to think about what you will put your harvest into when you get back to the skiff or car. A large tote will protect your car from drippings and protect your seaweed from potential contamination from boat gas and oil.

Check out possible beaches before your planned harvest. Check the beach for ease of access by boat or hiking. Assess the quality of the water—are there any outfalls, old industrial sites, logging dumps, etc., in the immediate area that would contaminate the seaweed?

Some seaweeds grow better in areas where there is high wave action; *Macrocystis*, *Porphyra*, and *Nereocystis* are good examples. Other seaweeds such as *Ulva* and *Palmaria* seem to grow more abundantly in sheltered bays and inlets. There are exceptions to this, but it will give you an idea of what to expect from a beach as influenced by its exposure to the open ocean.

It is advisable to get to the beach at least an hour before low tide. This will give you time to settle in and scout the beach for seaweeds and access routes to the water's edge. Remember that the rocks will likely be slippery so it is best to move slowly and deliberately. Be sure your knife or scissors are packed so they won't inadvertently impale you if you take an unexpected fall.

Harvesting

When you head to the shore you may be interested in harvesting only one type of seaweed such as *Palmaria* (ribbon seaweed) or you may want to harvest a variety depending on what you find. If you plan on harvesting more than one seaweed, it is advisable to carry several bags so you can place seaweeds into separate containers.

When you are collecting, do not take all the seaweeds in one area. Selectively cut, or "thin" seaweeds. Rock should not be left bare of seaweeds, or covered with cut stipes, or you will be destroying important habitat. Few seaweeds will regenerate, or grow back, from the stipe. Usually the holdfast is encrusted with small-shelled animals and is often tough and unpalatable. Leave the lower portion of the frond and holdfast to provide continued habitat for small animals.

It is best if you rinse the larger fronds at the ocean. This will help to remove small-shelled animals or any scum that may have settled on the seaweed from falling tides. Some seaweeds, such as *Porphyra* (black seaweed), need not be rinsed at all. More fragile seaweeds like *Ulva* (sea lettuce) may be rinsed at home with salted water.

Low tide reveals seaweeds in the rocky intertidal zone, near Valdez.

Processing

Most seaweeds can be dried and used throughout the year. *Fucus* (popweed) is an exception in that it loses much of its taste value once dried. Other seaweeds, like the bladelets of *Nereocystis* (bull kelp) or *Laminaria*, when dried concentrate their sugar and taste rather sweet. They do not taste sweet when fresh, however.

The large brown seaweeds can be harvested in volume and dried for winter use. To dry them, hang the seaweeds from a clothesline. They can be dried outside in the sun and in a nice breeze or inside with a fan if you have a tolerance for the smell of the ocean in your house for the day.

As seaweeds dry you may see a powdery substance appear. The white powder is likely to be a salt or a sugar and is perfectly edible.

Once seaweeds are dried, they can be stored in airtight containers such as jars or sealable plastic bags and then stored in a cool, dark area. Dried seaweed should easily last one year.

Alaria marginata
Winged Kelp, Wakame

Alaria, in the brown seaweed group, is known as winged kelp due to the bunch of small "blades" at the base of the frond. It is called wakame in Japan. This large seaweed grows in the lower intertidal zone.

Description
Alaria, an olive-brown colored seaweed, can grow up to more than 2 feet long and 2-8 inches wide in Alaska. It is anchored to the ocean floor with a visible holdfast. A short but noticeable

Alaria marginata (Winged kelp)
Color: olive-brown
Size: 2-8 inches x 2 ft
Collecting season: spring to early summer
Zone: lower intertidal

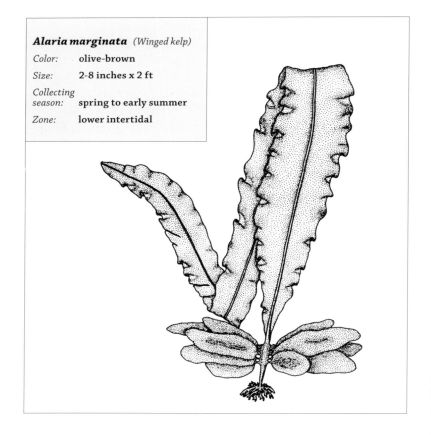

stipe runs from the holdfast to the frond. Oval shaped "blades" are attached to the stipe; these are sporophylls. The frond is long and narrow. The edges of the frond are ruffled and tend to fray near the tip, likely due to wave action. A thick midrib runs the length of the frond. Small *Alaria* fronds look similar to older ones.

Habitat

Alaria, like many larger brown seaweeds, tends to grow in the lower intertidal zone. It grows in areas with active surging water. *Alaria* often grows in large patches and is found attached to rock walls.

Harvesting

Alaria is a favorite for some people—the frond and midrib are delicious. Care should be taken not to overharvest an area. Generally this seaweed is harvested in late spring and into the early summer. Summer storms may batter the fronds until there is nothing left but a holdfast and a stub of a midrib.

Look for good-sized fronds and cut off only the top portion, leaving the lower portion of the frond and the stipe with the oval-shaped blades. You should be leaving a portion of every frond you harvest to maintain the habitat. Rinse in seawater or salted freshwater.

Harvested fronds can be kept in plastic or mesh bags until you get home.

Processing

Alaria will spoil fast if left in plastic for too long so begin processing right away when you get home. Fronds can be used fresh or dried. To dry fronds, hang them from a clothesline in the sun, if possible. As the frond dries it will shrivel up a bit.

Rather than drying the entire frond, many people slice the frond lengthwise into three strips—the midrib and two flat side pieces. The pieces can be hung to dry or used fresh.

Uses

When fresh the midrib of *Alaria* is often cut out, creating two sheets and a thick celery-like stalk. The midrib can be chopped fresh and quickly stir-fried. The two side sheets can be used like a tortilla. Rice and fish are piled in the center and the seaweed sheet is rolled.

Dried fronds or strips may be broken into pieces. Store the pieces in airtight containers, in a cool, dry place.

Alaria marginata (winged kelp) grows in the lower intertidal zone.

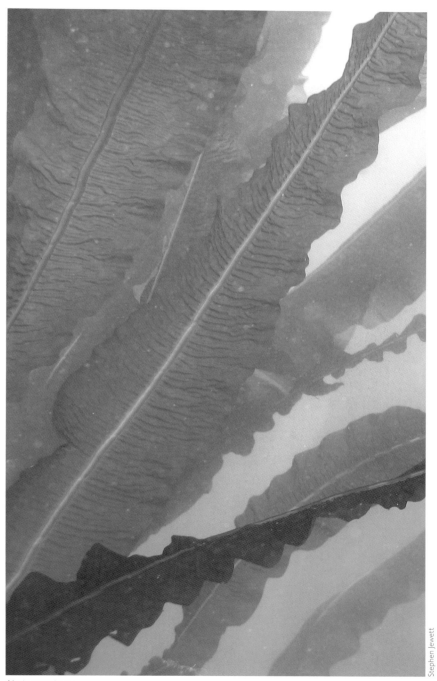

Alaria marginata

Fucus gardneri
Popweed, Rockweed

Fucus gardneri is in the brown seaweed group. It is often overlooked but is a good edible seaweed.

Description
Fucus, or popweed, is brown or yellow-brown in color, and is attached to rocks with a visible holdfast. There is no visible stipe. A thick stock branches into numerous bladelets.

Popweed has a visible midline that runs down the center of the frond as well as down the center of the branching bladelets. Mature popweed generally has bulbs or small sacs at branch tips. The surface of these small bulbs is textured or bumpy.

Popweed may grow up to 12 inches long in Alaska. The young specimens are shorter, 2-4 inches, and appear more yellowish and often don't have bulbs on branch tips. These younger ones are preferred for picking.

Habitat
Popweed is most abundantly found in the upper intertidal zone; you will find it on almost any low tide. It grows along rocky shorelines, attached to solid substrates such as large and smaller rocks.

Harvesting
Popweed can be picked any time of the year, although younger yellowish specimens are more abundant in the late spring or early summer. You will find popweed on many rocky shores along the Gulf of Alaska and into Bristol Bay.

Because popweed is not as tasty after it has been dried, it is best to pick only as much as you will eat in the next few days.

While the younger yellow-brown popweed is preferred, the larger, darker brown forms are also tasty.

Fucus gardneri

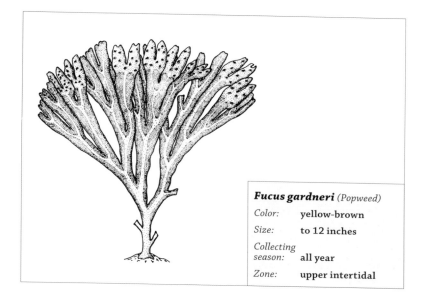

Fucus gardneri (Popweed)
Color: yellow-brown
Size: to 12 inches
Collecting season: all year
Zone: upper intertidal

For the younger popweed, take the frond but be careful not to tear away the holdfast. With larger popweed, you do not want to pick the entire frond. Instead, you should harvest only the branched bladelets that do not have bulbs on them. The branched bladelets can be cut or torn away from the main frond.

Processing

As mentioned earlier, popweed tastes best fresh. Rinse it in ocean or freshwater. While it can be eaten raw, it has a "nuttier" flavor if blanched by dipping into boiling water.

Uses

There are several recipes for this versatile and abundant seaweed.

It is fun to introduce kids to seaweeds by blanching popweed for them on a beach outing. Take along a thermos of hot water, a small pot or coffee can, and a small paring knife or scissors. Be prepared to make a small beach fire.

Look for the fronds that do not have bulbs on the tips. The water inside these bulbs is a bit slimy and grosses kids out.

Pull off some fronds near the base. Pour the thermos water into your can over the fire. Have kids watch you while you dip the seaweed into the hot water for a few seconds, holding it by the base. The normally olive-brown color will turn a bright, near-neon green. For some reason kids think this makes it okay to eat. The water needs to be hot—if the water is not hot enough, the neon green color change may not occur.

The younger specimens of *Fucus gardneri* (popweed) are preferred for picking.

Mature popweed, with bulbs at the branch tips, is tasty but not as desirable as the young ones.

Laminaria
Kelp, Kombu, Sugar Wrack

Laminaria is in the brown seaweed group. Often called kelp or kombu, it is abundant in Alaska's gulf waters. Several species of *Laminaria* are large, ranging in size from 1 to 4 feet when mature. Common *Laminaria* in the Gulf of Alaska include *Laminaria saccharina* (commonly called sugar wrack), *Laminaria bongardiana* (formerly *Laminaria groenlandica*), and *Laminaria setchellii*. There are other species of *Laminaria* in Alaska's waters too.

Description
All species of *Laminaria* have large visible holdfasts. Beyond that, their shape varies considerably. Some have short stipes and long blades such as *Laminaria saccharina*, or longer stipes and large blades as in *Laminaria bongardiana* and *Laminaria setchellii*. In all cases the frond or blade is thick and dark brown. Another similarity among *Laminaria* species is the lack of a midrib.

The shape of the blade itself varies considerably. The blades of *Laminaria saccharina* have large bumps or indentations. *Laminaria bongardiana* has a thick smooth surface and the blade is split into several larger strips. *Laminaria setchellii* has numerous long, relatively thin blades all running from the same base.

Habitat
Kelps are found in lower intertidal waters. You may find a rock covered with one type of *Laminaria* or you may find several different species in one bay or inlet. The thicker bladed *Laminaria bongardiana* and *Laminaria setchellii* are found in more exposed areas and the thinner bladed *Laminaria saccharina* is seen in areas with less wave action. All three of these seaweeds have been found in one area, such as around Craig or Sitka.

Laminaria *(Kelp)*

Color:	browns
Size:	about 1-4 ft
Collecting season:	spring to fall
Zone:	lower intertidal

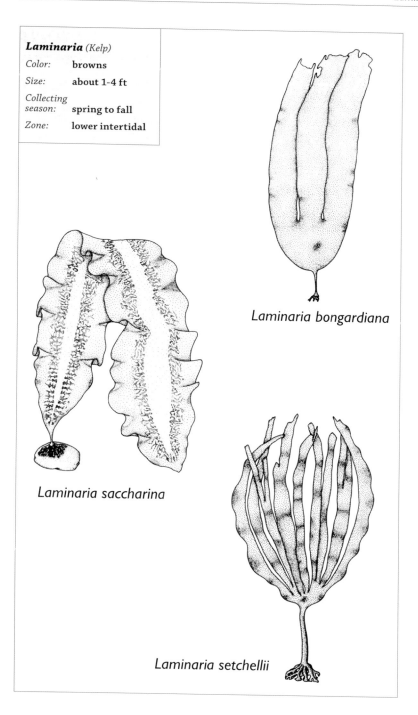

Laminaria bongardiana

Laminaria saccharina

Laminaria setchellii

Harvesting

Kelps can be harvested from late spring into autumn. As the season progresses they may get beat up from wave action or may be settled on by small invertebrates or sponges. Begin looking for these large seaweeds in April and May. They generally require a "minus" tide to harvest.

These large seaweeds are easy to pick. Prepare for harvesting by bringing a couple of plastic or mesh sacks, old pillowcases, and a small paring knife. As with other seaweeds, do not pick the entire kelp patch. Their blades provide a canopy for small fish and invertebrates and are important to the health of the local ecosystem.

Check the seaweed to see that it looks healthy with no sponge encrustations. Look for fronds that don't have ragged edges or patches of lighter coloration which signify damaged or diseased parts.

Taking only a few plants from one area, use a paring knife to slice off the blade leaving the holdfast, stipe, and lower portion

The thinner bladed *Laminaria saccharina* (kelp) is found in areas with less wave action.

of the blade. Rinse the blades in ocean water to remove small crustaceans or sand. If possible, put different species in different bags for transport back home.

The thicker bladed *Laminaria bongardiana* is more common in exposed areas.

Processing

Once you get home, there is little work to processing. To dry, string up twine in an area where there will be some sun and wind. The blades can be hung from the twine, and can be attached with clothespins or simply draped over the twine.

As blades dry they will turn a very dark brown, then blackish. A dusting of white may appear on the surface. This is salt, or with *Laminaria saccharina*, it is the sweet-tasting mannitol, and is edible.

Kelp, like most other seaweeds, is stored in an airtight container in a cool dry area. Large pieces can be broken into smaller pieces for storing.

Uses

As you begin to experiment with these seaweeds you will find that each of the kelps has a different flavor and texture. *Laminaria setchellii* and *Laminaria saccharina* are both sweet when dried. *Laminaria bongardiana* dries thicker, and when dried pieces are soaked in freshwater for 30 minutes they will appear fresh, as if never dried. See the recipe section for kelp chips, kelp seasoning, and other dishes.

Nereocystis luetkeana
Bull Kelp, Bullwhip Kelp

Nereocystis luetkeana is a common seaweed found along the Pacific west coast. Many people are familiar with bull kelp, but do not know it is edible.

Description
Nereocystis (bull kelp) is one of the largest seaweeds in the North Pacific, growing up to 100 feet in length. Bull kelp, in the brown algae group, is golden to dark brown in color.

Bull kelp is attached to the ocean bottom with a stout root-like holdfast, off of which grows a long hollow cylindrical stipe or stalk. The stipe is narrow at the base and increases in diameter farther from the base. The stipe is topped with a hollow bulb, a pneumatocyst, to which are attached numerous long leaf-like blades.

Habitat
Bull kelp grows in subtidal areas and usually we see only the top portion—the bulb and blades. Rarely would an attached holdfast be exposed at a low tide, although you may find holdfasts and portions of the stipe, bulb, or blades washed up after a good storm. Bull kelp grows along many coastlines, more abundantly in areas having high wave action. Bull kelp is disliked by small boaters who consider it a nuisance when trying to motor into a small cove or through a narrow pass.

Harvesting
Most of the harvesting occurs in the summer and early fall, although specimens can be harvested year-round if they are healthy. You will likely not be able to reach bull kelp at a low tide on foot, as its holdfast is attached to rocks in the subtidal area. If you find some fresh bull kelp washed up along the shore

after a storm, it could be okay to collect and use if it is still olive brown and appears moist.

Bull kelp is usually harvested from a skiff. A long knife and a bag are necessary for harvest. Idle your skiff up to the bull kelp and grasp the long stipe or stem. Use your knife and cut down as far into the water as you can—and don't fall in! Or you can harvest the long blades from the top of the bulb. Both the stipe and blades are good to eat.

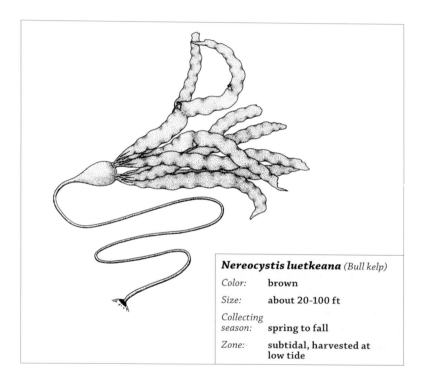

Nereocystis luetkeana (Bull kelp)

Color:	brown
Size:	about 20-100 ft
Collecting season:	spring to fall
Zone:	subtidal, harvested at low tide

Processing

The stipe is used differently from the blades. The stipe should be rinsed in freshwater to remove excess salt. The hollow stipe is most often used in pickling and not dried before being pickled.

The blades can be dried or used fresh. To dry, rinse quickly in freshwater. Hang to dry on a clothesline, in a slight breeze or sunshine. The blades can be dried inside if it is raining. As blades dry, your house will smell like the sea and the blades will turn a darker brown and shrivel up a bit. Once dried, the blades can be kept in airtight containers in a cool dry place.

Uses

A variety of pickles or relishes can be made from the thicker portion of the stipe/stem. You can experiment by using pickle or relish recipes in a standard cookbook and replacing the cucumber with bull kelp.

The long-stiped *Nereocystis luetkeana* (bull kelp) is usually harvested from a skiff.

Nereocystis luetkeana

Nereocystis luetkeana (bull kelp)

Porphyra
Black Seaweed, Nori, Laver

Porphyra is in the red seaweed group. Several species of *Porphyra* are found along the west coast and approximately 30 species exist worldwide. Locally it is called black seaweed.

Description

Black seaweed is an annual—it grows and dies back each year. Black seaweed begins to grow in early spring. It is recognizable by the near black strands hanging down rock faces. The near transparent fronds may be a dark rose-purple or a black-green color. The colors are most apparent when the fronds are wet; black seaweed appears nearly black when drying.

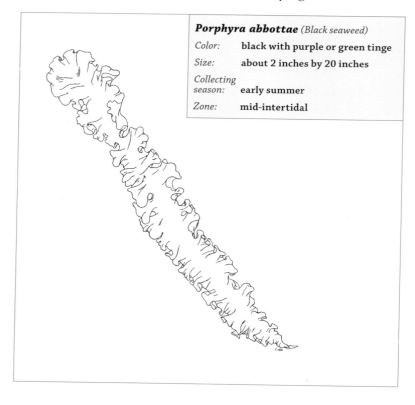

Porphyra abbottae (Black seaweed)
Color:	black with purple or green tinge
Size:	about 2 inches by 20 inches
Collecting season:	early summer
Zone:	mid-intertidal

Several species of *Porphyra* occur in Southeast Alaska waters. All are edible and nutritious. In all species there is no apparent midrib or stipe and only a small holdfast. The edges of the fronds are ruffled.

Porphyra abbottae is the preferred species for harvesting. Individual fronds can grow up to 20 inches long and about 2 inches wide along the length of the frond. *Porphyra perforata* grows in a circular shape up to 12 inches in diameter, with the frond anchored by a holdfast which originates near the middle of the frond. You may find these two species in the same area or you may find only one type when you go out to harvest. *Porphyra torta* is the "earlier" or winter black seaweed, and is common around Ketchikan, Klawock, Craig, Sitka, and other areas. A fourth small species, *Porphyra nereocystis*, which usually is not found while on a seaweed picking outing, grows attached to the stipe of *Nereocystis* or bull kelp.

Habitat

Porphyra abbottae is commonly found in the outer coasts along the Southeast Alaska panhandle and less commonly in the inside waters of the Panhandle. It is found around mid- to lower intertidal ranges in areas with high wave action. When this seaweed begins growing in the spring it will often take over an entire rock. In the North Pacific, *Porphyra abbottae* begins growing in the early spring, reaches its peak in late spring, and dies off over the summer.

The other local *Porphyra* species have slightly different growing seasons and habitat preferences. *Porphyra torta*, locally called winter black seaweed, can be found in late winter and into early spring, although in smaller quantities.

The lower-intertidal *Porphyra abbottae* (black seaweed) will often take over an entire rock.

Two species of black seaweed, *Porphyra perforata* and *Porphyra abbottae*, are often found growing right next to one another. The long, narrow *P. abbottae* is preferred for harvesting.

Harvesting

Timing is important when getting ready to harvest black seaweed. If you go out too early the seaweed will be short and hard to pick. If you are too late it may be encrusted with small snails or already dying back and turning a light brown. The time to pick can vary each year with changes in temperature, sun, and rain. The sun can speed the growth so you may pick earlier in the season during a sunny spring. Rain can wash the thin frond of its flavor so don't pick right after a rain. These variables require that you pay attention to the weather. You may need to do some scouting ahead of time and go out several times in your skiff to see if the seaweed is ready to pick. Early May is a benchmark for harvesting; however, it can be earlier or later depending on variables.

Test for readiness by picking a frond and stretching it along its length. It should be elastic and stretch. It can be 8-15 inches long. If it is too short, it will be hard to pick. By the time it gets longer small snails may have settled onto the fronds. Later in the

When *Porphyra abbottae* is 8-15 inches long and somewhat elastic, it's ready for picking.

summer, *Porphyra abbottae* turns a light brown as it begins to die away.

In your skiff be sure to take the necessary safety gear, which includes a life jacket and a hand-held radio, along with a pillowcase or mesh bag for collecting.

Pick by simply pulling on each frond or small group of fronds. They will come off the substrate easily. Your knuckles may get beat up after awhile, but the harvest is well worth it. Black seaweed grows in large quantities and is easy to harvest; therefore it can be easy to overharvest. Remember, the more you harvest the more processing work you will have to do after you get home.

A number of people harvest black seaweed, so the resource is well used. While black seaweed seems to be in high abundances in a few west coast locations, most of these sites are picked by several communities and numerous families.

Processing

Black seaweed is delicate and several steps must be taken to ensure a quality product. This delicate frond should **not** be rinsed in freshwater. If it is rinsed in freshwater, it will lose most of its taste.

The harvest should be laid out flat to dry on an old bed sheet. It can be put outside on a makeshift plywood table.

Author Dolly Garza demonstrates drying black seaweed. The piles need to be pulled apart several times during the drying process.

A normal table height allows you to comfortably work on the seaweed.

Expose the seaweed to both a light breeze and sunlight. If you don't have these optimum conditions, try for a nice breeze or at least keep it out of the rain. The more radiant heat you have, the quicker the drying.

The many individual fronds will tend to stick together like glue as they dry, forming irregularly shaped balls. If the seaweed is allowed to dry like this the center will remain wet and eventually mold while the outside appears dry.

You need to pull these little piles of drying seaweed apart several times over the course of the day to ensure even drying. This often takes hours. The seaweed should also be turned over as you are pulling it apart. On a good day the seaweed can be dried in a 12-hour period. In damp weather it often takes two days. Bring the sheet in overnight, and then put it back out the next day. After it feels dry, the seaweed can be finished by oven roasting at around 175° for 10-20 minutes. The final oven roasting ensures that the moisture is removed. The seaweed can then be stored in an airtight container in a cool dry place.

Uses

Black seaweed is a prized food to Tlingit, Haida, and Tsimshian. It is an important trade item, because there are many areas where it does not grow, and because many Elders who enjoy eating it have stopped harvesting black seaweed because it is too much work to harvest. Members of non-Native rural communities also enjoy harvesting black seaweed in the spring.

Generally, black seaweed is dried and not used fresh. Dried black seaweed is eaten as a snack like popcorn. It can also be added to a meal like the Seaweed Chop Suey described in the recipe section.

Black seaweed surfaces need to be exposed to the air, to aid the drying process.

Palmaria mollis
Ribbon Seaweed, Dulse

Palmaria is called ribbon seaweed in Southeast and Southcentral Alaska, and dulse in the Lower 48 states. It is one of the red seaweeds, in the group Rhodophyta. There are several species in the Gulf of Alaska region including *Palmaria mollis*, *Palmaria hecatensis*, and *Palmaria callophylloides*. While there are similarities in these three species, this discussion is specific to *Palmaria mollis*.

Description
Ribbon seaweed is brick red or reddish-brown in color. Several blades branch off a single small holdfast. There is no stipe or stem. Fronds sometimes appear like lobes while others may appear more irregular in shape. They grow longer than wide, up to 8-10 inches long and several inches wide. The blades are thick and almost feel leathery to the touch.

Palmaria mollis (ribbon seaweed) grows in the upper intertidal zone.

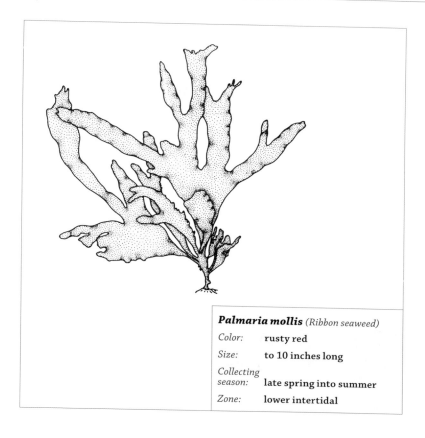

Palmaria mollis (Ribbon seaweed)
Color: rusty red
Size: to 10 inches long
Collecting season: late spring into summer
Zone: lower intertidal

Habitat

Ribbon seaweed grows in a variety of habitats from sheltered bays to exposed rocks along the west coast. It grows in the lower intertidal zone at the −1.5 to −2 foot range. You will find clumps of it growing here and there, but it will not take over an area like black seaweed (*Porphyra*) or some of the browns (*Laminaria*) do. You will find ribbon seaweed in the late spring and summer and sometimes in the early fall.

Harvesting

Ribbon seaweed can be harvested from late spring into summer depending on weather and tidal action. Harvest by cutting or carefully tearing away the blades from the holdfast. Be sure not to harvest all of the fronds in a small area, but leave some to provide shelter to small-shelled animals. Rinse the fronds in seawater.

Processing

Ribbon seaweed will dry to a leathery texture and will remain slightly pliable even when dried. Dry by laying it on a bed sheet in the sun, exposed to a light breeze. You may see columns or pillows of white forming on the surface of the fronds as they dry. This is salt and is edible. Some people dip the seaweed in a sugar water to prevent these deposits. Also, if you dry the seaweed quickly in sun, the salt pillows may not occur. Simply wipe off the salt formations with a slightly damp cloth, finish drying, and then store.

Uses

Dried ribbon seaweed has a tough, chewy texture. Some people prefer it cooked in a dish, or fried or roasted like a chip.

Palmaria mollis

Palmaria mollis (ribbon seaweed or dulse)

Ulva fenestrata
Sea Lettuce

Ulva fenestrata is commonly called sea lettuce. It is one of the few large, bright green seaweeds in the group Chlorophyta found in the Gulf of Alaska. It is a delicate seaweed with a mild flavor.

Description

Sea lettuce is vivid green and cellophane thin. You can see the outline of your hand even through larger mature fronds. The frond is connected to rocks with a small, almost invisible holdfast. The frond grows as a single, irregular, but somewhat round shaped blade with slightly ruffled edges. The young fronds are small, thin, and very fragile, while the large fronds can grow to 6-8 inches and feel slightly thicker. There may be randomly spaced holes in the frond.

Sea lettuce can be confused with other green seaweeds which may not have the same flavor as *Ulva*.

Ulva fenestrata (sea lettuce) is a delicate seaweed, abundant in sheltered bays.

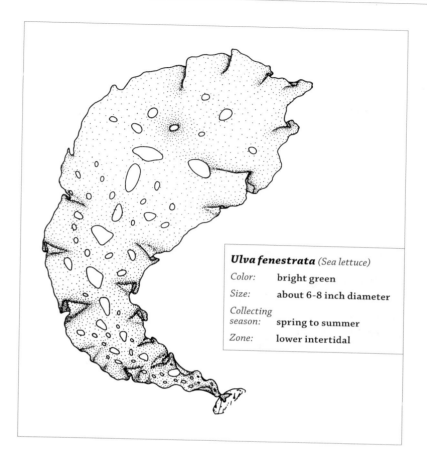

Ulva fenestrata (Sea lettuce)
Color: bright green
Size: about 6-8 inch diameter
Collecting season: spring to summer
Zone: lower intertidal

Habitat

Sea lettuce can be found in an array of habitats, but is seen more abundantly in sheltered bays or in areas with limited wave action. It grows in the lower intertidal, in the −1.5 to −2 foot tidal range.

Ulva and other green algae grow well in areas polluted with sewage, since they are nutrient scavengers. Be sure you are not collecting near a septic tank drain field or other polluted area.

Harvesting

Sea lettuce can be picked from spring into the summer depending upon weather and amount of sun. Harsh weather can batter it, leaving it too tattered for picking.

Sea lettuce often does not grow in large patches, so harvesting yields small amounts. Harvest with scissors or a small knife and carefully cut the blade from the holdfast. If the holdfast is accidentally pulled off, cut the holdfast from the frond before processing. Quickly rinse the sea lettuce in seawater to remove small animals.

Processing

The blades are small and thin and are best dried by laying them on an old bed sheet in the sun. A light breeze will help. As the blades dry, they will shrivel some and turn a darker green.

Uses

Sea lettuce is usually picked to dry and used as a seasoning. It is a delicate seaweed that dries to a small amount.

Sea lettuce shrivels and turns darker when it dries.

Salicornia virginica
Beach Asparagus

Description

Beach asparagus, *Salicornia virginica*, is a flowering plant in the goosefoot family, Chenopodiaceae. It is not a seaweed, but can be collected easily during expeditions to harvest seaweed.

Beach asparagus grows along the Pacific Northwest coast and the Atlantic coast, and a similar species (*Salicornia europaea*) is widespread in Europe. Beach asparagus is also known as sea asparagus, pickleweed, saltwort, and glasswort.

Beach asparagus is a native succulent perennial plant, growing up to around twelve inches in height. The stems are scaly and jointed. When mature, it has minute clustered greenish flowers at the tips of the stems. The tiny flowers turn a purple-red as they mature, causing the stems to slowly gain a light red tinge toward the tips.

Habitat

Beach asparagus is found along seashores, in the high intertidal area where the plant is covered by the ocean at hide tides and exposed as the tide falls. Beach asparagus is generally found in sandy bays and along protected shorelines. It can develop large mats in favorable conditions.

When found along more rocky shores the plants may be more sparsely scattered. It grows more abundantly in southern Southeast Alaska and in protected bays. As you travel north it can be a little tricky to find, but it is worth the effort.

The plant dies back each year, then begins to grow again in the late spring in the same spot. By mid-June you will begin to see new growth. The plant grows quickly and will be ready to pick by the end of June or in early July. As the plant continues to

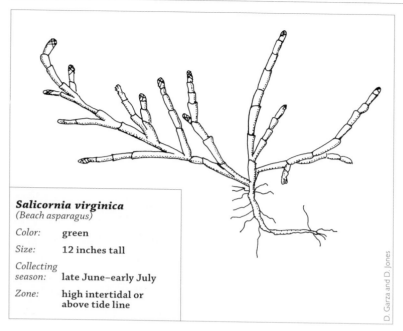

Salicornia virginica
(Beach asparagus)
Color: green
Size: 12 inches tall
Collecting season: late June–early July
Zone: high intertidal or above tide line

grow and begins flowering it becomes more "woody" and is less desirable for picking.

Harvesting

When preparing to pick, there are several considerations. If you are new to an area check with locals to find out where beach asparagus grows. Keep in mind that easy-to-pick spots are often harvested by older ladies who still enjoy this early summer activity. Don't compete with them as they probably don't have the strength or means to go to more distant locations.

In most cases you will go by skiff to find asparagus. Spend time before early July to locate a bay with a good mat of asparagus.

To pick beach asparagus you need a small paring knife and a pillowcase or large plastic bag (not a garbage bag).

Harvest beach asparagus when the stems are around 10 inches tall, filled out a bit (not too skinny), but not yet flowering (the reddish tips at the top of the stalks). Like most plants there is a short window of opportunity for picking. If you pick too early the stems are skinny and short. If you pick too late and they've started flowering the stems will be woody and not as tasty.

To harvest, sit or kneel on the beach. Grab a small hunk of stems with one hand and use the paring knife to slice the stems off near the base of the plant.

Do not pull the plant out. If you do, it will be a lot of work to clean later. The trick is to pick clean. With every handful that you sever from the base of the plant, take five seconds and pick out the grass and other debris (such as dried seaweed). If you don't remove the debris on the beach you will have a lot more work to do when you get home.

Processing

Once you get home the work starts. You need to rinse your batch in cold water and pick out any of the debris you missed. If you are a clean picker this could take a half-hour. If not, it could take hours.

Beach asparagus will last for at least a week in your refrigerator; if you can, simply eat your harvest over the course of a week. Or you can freeze or can beach asparagus.

To prepare asparagus for freezing, start with blanching. Use a colander or large slotted spoon to put a bunch of asparagus into softly boiling water. Leave it in for about 30 seconds, then take out. Place in a large colander to drain excess water, then rinse with cold water. To freeze, place blanched and drained spears in plastic freezer bags, squeeze excess air out, label, and freeze.

To can beach asparagus, follow instructions for canning string beans.

Uses

Beach asparagus is eaten as a vegetable, similar to asparagus or green beans. See recipes for beach asparagus in the recipes section.

Recipes

Seasonings

Kelp Seasoning

Kelp powder or flakes can be made by roasting *Laminaria* blades in an oven at low temperatures, 150-200° for 5 to 15 minutes depending on what seaweed you are roasting and how much you have in the oven. They will turn a bright green and become quite crisp. Remove from the oven. A powder or flakes can be made by crushing with a mortar and pestle or crumbling in your hands. Used as a seasoning, it can be added to seafood chowder or sprinkled over rice.

Sea Lettuce Flakes

Roast dried blades of *Ulva* in an oven at 180-200° for 5-15 minutes or until the blades are just crisp. They will easily crush to a powder or small flakes in your hand or with a mortar. Store in a jar and use as a seasoning in soups or on salads.

Kombu (Kelp) Stock

Kelp (kombu, *Laminaria bongardiana)* can be added to water to provide a flavorful stock base for soups. The kelp is not boiled but put into water and heated. It is then removed from the water just before boiling. The stock can then be used to make your soup or rice dish.

Snacks

Quick Fry Kelp Chips

Cut fresh *Laminaria saccharina* or *Laminaria setchellii* into small chip-like strips. Heat a frying pan with a few tablespoons or more of oil. Quickly turn strips until green. Remove and transfer to paper towel to remove excess oil. You may add a bit of sugar or sesame seeds.

Oven Roasted Kelp Chips

Roast *Laminaria saccharina* or *Laminaria setchellii* strips in an oven at 200° for 5 to 10 minutes, or until seaweeds turn green. Remove, cool, and eat.

Ribbon Seaweed Chips

Like many of the other seaweeds, ribbon seaweed (*Palmaria*) is delicious when roasted. Put some in a shallow pan and roast it at 125-175° for about 5 minutes. Remove from oven and allow to cool. It will crisp up as it cools. It has its own salt and flavor so there is no need to add seasoning.

Bull Kelp Chips

Break dried bull kelp (*Nereocystis*) blades into smaller pieces. Put in shallow pan in an oven at 125-175° for 5-10 minutes. Test for "doneness" by breaking off a piece. If it breaks with a crunch, it is done. The chips should be salty enough from the natural salts and not require any additional seasoning. This seaweed chip has been favored by many participants at seaweed workshops.

Roasted Seaweed Popcorn

Fill a shallow pan with a single layer of black seaweed (*Porphyra*). Roast in an oven at around 175° for approximately 10 minutes. Check to see if it is roasted by trying to bend and snap a piece. If it snaps, it is ready to munch on like popcorn. Roasted seaweed may act as a laxative if you eat too much at once, so be a bit cautious.

Main Dishes

Alaria Tomato Noodle Dish

⅓ pound ground beef
2 tablespoons oil
1 small onion, chopped
2 cloves garlic, sliced
2 cups sliced *Alaria*, fresh or rehydrated dried*
1 (8 ounce) can tomato sauce
1 teaspoon vinegar
½ teaspoon oregano
½ teaspoon marjoram
½ teaspoon basil
⅛ teaspoon cayenne
Ramen noodles, cooked and drained

Brown meat in oil with onion and garlic, then drain off most of the fat. Add remaining ingredients except noodles. Cook 10 to 15 minutes and serve over cooked ramen noodles.

To rehydrate dried Alaria, soak in freshwater for about 15 minutes.

Stir-Fry Veggies with *Alaria*

1 package oriental noodles
1 tablespoon soy sauce
1 tablespoon sesame oil
2 teaspoons powdered garlic
2 tablespoons cooking oil
2 carrots sliced on
  the diagonal
¼ medium cabbage,
  sliced or shredded
1 handful of sugar peas,
  sliced in half
½ cup fresh or rehydrated
  Alaria, cut into
  small strips
1 tablespoon soy sauce
2 teaspoons honey

Cook noodles according to package directions. Drain and flavor with 1 tablespoon soy sauce, sesame oil, and garlic powder. Put into a large bowl. Stir-fry vegetables, except for *Alaria*, in cooking oil. Add *Alaria* when the vegetables are half done. Add 1 tablespoon soy sauce and the honey to the vegetables while stir-frying. Add vegetables to the noodles and toss. Serve warm.

Fish Chowder with Rockweed

2 large potatoes, diced
2 quarts water
2 cups rockweed
  (*Fucus*), chopped
1 pound rockfish
  fillet, chunked
1 cup chicken broth
1 tablespoon celery salt
½ teaspoon sage
½ teaspoon thyme
⅛ teaspoon cayenne
½ teaspoon salt
2 tablespoons olive oil
1 tablespoon flour

Place potatoes in soup pot with water. Put chopped rockweed in cheesecloth and suspend the bag in the soup pot. Boil until potatoes are almost cooked. Remove the spent rockweed and discard. Add all other ingredients except flour. Lightly boil for 15 minutes to cook the rockfish. Dissolve flour in a few tablespoons of cold water, and stir it into chowder to thicken. Simmer five more minutes, then serve.

Fucus Stir-Fry

3 tablespoons oil
1 garlic clove, grated
2 tablespoons ginger, grated
½ cup mushrooms, sliced
2 cups fresh rockweed
    (*Fucus*), cut into pieces*
½ cup celery, sliced

½ cup water chestnuts, sliced
½ onion, chopped
¼ cup red pepper, sliced
1 cup edible pea pods
3-4 tablespoons soy sauce
1 teaspoon chicken broth
Cooked rice or cooked
    noodles

Heat oil in skillet or wok. Sauté garlic and ginger. Add in order: mushrooms and rockweed, then celery, water chestnuts, onions, pepper, and peas. Add soy sauce and broth. Cover and steam briefly to allow flavors to blend. Serve with rice or noodles.

You can also use Alaria *or kelp* (Laminaria).

Fucus Chop Suey

1 teaspoon cornstarch
1 teaspoon chicken broth
1 teaspoon soy sauce
1 cup pork, chicken, or
    tofu (thinly sliced)
3 tablespoons oil
1 garlic clove, grated
2 tablespoons ginger, grated
2 cups fresh rockweed
    (*Fucus*)*

½ cup mushrooms, sliced
½ cup celery, sliced
½ cup red pepper, sliced
1 cup edible pea pods
½ onion, chopped
3-4 tablespoons soy sauce
1 teaspoon chicken broth
Cooked rice or cooked
    noodles

Combine cornstarch, 1 teaspoon chicken broth, and 1 teaspoon soy sauce, and marinate meat in mixture for a few minutes. Heat wok and add oil, meat, garlic, and ginger. When meat is cooked, remove from wok and set aside. Sauté vegetables, including rockweed, in wok. Return meat to the wok, add soy sauce and chicken broth, and simmer briefly to allow flavors to blend. Serve with rice or noodles.

You can also use Alaria *or kelp* (Laminaria).

Snapper and Tofu

1 strip kelp *(Laminaria)* about 6 inches long
½ cake firm tofu, cut into 1-inch cubes
1 pound red snapper* fillet or other rockfish, cubed
1 lemon rind, cut into thin strips

3 tablespoons soy sauce or tamari
Juice of one lemon
2 tablespoons honey
Cooked rice

Place kelp in the bottom of steamer pot. Cover with water and soak while preparing other ingredients. Place tofu and fish in steamer basket with lemon rind strips. Cover and steam for 15 minutes or until rockfish fillet is done. Remove from heat. In a small saucepan mix soy sauce, lemon juice, and honey. Heat only until honey melts. Arrange the steamed fish, tofu, and lemon peel on a platter. Pour the sauce over it and serve hot with rice. Discard kelp.

* *"Red snapper" is a common name for yelloweye rockfish.*

Dulse Fried Rice

1 tablespoon cooking oil
½ cup carrots, finely chopped
½ cup green onion, chopped
1 handful dried ribbon seaweed (dulse), snipped into small pieces

3 cups cooked brown rice
Soy sauce to taste
1 egg, scrambled (or tofu)

Heat oil in frying pan, then add carrots and stir for about 30 seconds. Add onion and ribbon seaweed, and stir. Add rice, breaking up clumps, and stir until hot. Add soy sauce to taste and stir. Fold in eggs or tofu. Serve warm.

Nori with Bean Threads

1 cup bean threads
1 tablespoon sesame oil
½ cup carrots, sliced
½ cup onion, sliced
½ cup mushrooms, sliced

1 cup roasted nori
  (black seaweed)
½ cup water
Soy sauce
Sesame oil
Cooked rice

Cover bean threads with boiling water, then set aside for five minutes. Drain and cut threads in half. Sauté vegetables in oil. Add black seaweed when vegetables are almost done. Add bean threads and ½ cup of water and cook for two minutes. Season with soy sauce to taste. Sprinkle with sesame oil just before serving. Serve over rice.

Seaweed Chop Suey

½ pound bacon, cut
  in thin strips
2 garlic cloves, peeled
  and sliced
½ large onion, sliced
1 medium cabbage, sliced
¼ cup soy sauce

1 can chopped canned
  clams, drained (save
  and set aside juice)
1-2 cups dried black
  seaweed, ground
Cooked rice

Fry bacon until crisp. Add garlic and onion and sauté for half a minute. Remove half the drippings. Add cabbage, soy sauce, and clam juice. Allow cabbage to cook. Add clams and 1-2 cups of dried ground black seaweed. Mix and add more water if necessary. Serve over rice.

Black seaweed should be unfolded and turned throughout the day, as it dries.

Side Dishes

Sweet and Sour *Fucus*

3 tablespoons oil
4 tablespoons sunflower seeds
2 cups rockweed *(Fucus)*,
    cut into strips
2 cups tart apples,
    peeled and sliced
¼ cup raisins
1 teaspoon cinnamon
2 tablespoons honey

In skillet, sauté sunflower seeds in oil until slightly brown. Add rockweed and sauté until tender. Add apple slices, raisins, cinnamon, and honey. Turn heat down and simmer for a few minutes. Serve as a side dish.

Dolly Garza looks at the condition of *Porphyra* (black seaweed) to find out if it's still good for picking.

Seaweed Rice Balls

2 tablespoons honey
4 tablespoons soy sauce
2 tablespoons sesame oil
¼ teaspoon ground ginger
8 cups cooked rice
⅓ cup sesame seeds
2 cups roasted black seaweed
    (Porphyra), crushed

In small bowl mix honey, soy sauce, sesame oil, and ginger. In large bowl, mix rice with sesame seed and then the marinade. Add the crushed black seaweed (reserving a small amount to use as garnish). Form 1-inch rice balls. Place on platter and sprinkle lightly with crushed black seaweed. You can also serve as a rice dish without making the balls.

Sautéed Beach Asparagus

2-3 cups fresh beach
    asparagus
¼ cup small fried bacon bits
    (3 slices raw bacon)
2 tablespoons butter
½ onion, sliced
Lemon juice to taste

Clean fresh beach asparagus, then soak for 5 minutes in cold water. Drain, then steam or blanch for 1-2 minutes. Rinse in cold water and drain. (Skip this if you are using canned beach asparagus.)

Chop bacon slices and fry. Remove and drain on paper towel. Drain frying pan, but do not wipe clean. Add butter and sauté onion for 1-2 minutes. Add asparagus and sauté for 1-2 minutes more. Turn mixture into bowl and sprinkle with lemon juice. Serve warm.

Black seaweed drying on tables

Beach Asparagus with Parmesan

2 cups beach asparagus, fresh or canned
2 tablespoons extra virgin olive oil
½ teaspoon lemon juice
½ cup parmesan, shredded
Black pepper, ground

If the asparagus is fresh, blanch for one minute in unsalted boiling water. If asparagus is canned, heat it in small pan for about 5 minutes. Drain and place in serving bowl. Make a dressing with olive oil and lemon juice, and spoon over the warm beach asparagus. Sprinkle with parmesan and ground black pepper. Serve warm.

Greek Beach Asparagus Salad

2-3 roma tomatoes, chopped and drained
1 cucumber, peeled, sliced, and chopped
½ small red onion, chopped
1 jar plain beach asparagus, drained
½ cup kalamata olives, chopped
½ cup feta cheese, crumbled
2 tablespoons olive oil
2 tablespoons balsamic vinegar

In bowl, toss tomatoes, cucumber, and onion. Add drained beach asparagus and toss to separate "spears." Add olives and feta, then toss. In a small bowl mix oil and vinegar, then add to salad and toss. Refrigerate before serving.

Sliced bull kelp.

Canned Products

Note: To ensure you are following procedures for safe food products, it is recommended that you read about food canning at the National Center for Home Food Preservation Web site, www.uga.edu/nchfp/index.html.

Bull Kelp Chutney I

3 cups bull kelp
    (*Nereocystis*), chopped
2 cups raisins
2 cups apples, chopped
2 cups rhubarb, chopped
¾ teaspoon salt

3 ½ cups brown sugar
1 pint cider vinegar
3 ounces mustard seed
½ teaspoon ginger
Cayenne to taste

Combine all ingredients in a large pot. Bring to boil, then simmer (uncovered) for about one hour. Place chutney in half-pint or pint sterilized jars, leaving ½ inch headroom. Screw on lids, and boil submerged in water for 15 minutes. Remove jars from hot water and let cool for 12-24 hours, then check lids to make sure they sealed.

Bull Kelp Chutney II

9-10 cups bull kelp
    (*Nereocystis*), chopped
1 ½ cups onion, chopped
2 cups raisins
½ cup ginger, chopped
5 cloves garlic
3 cups vinegar

2 ½ cups sugar
¾ teaspoon cayenne
    (or to taste)
½ teaspoon salt
½ teaspoon cloves, whole
½ teaspoon cinnamon
½ teaspoon allspice

Combine all ingredients in a large pot. Bring to boil, then simmer about two hours until syrupy. Place chutney in half-pint or pint sterilized jars, leaving ½ inch headroom. Screw on lids, and boil submerged in water for 15 minutes. Remove jars from hot water and let cool for 12-24 hours, then check lids to make sure they sealed.

Bull Kelp Salsa

8 cups bull kelp *(Nereocystis)*, chopped*
4 green peppers, chopped*
2 onions, chopped*
3 cups celery, diced*
5 large fresh tomatoes, chopped *
Red chili peppers to taste
1 green Ortega chili (or more to taste)
3 garlic cloves, crushed
½ teaspoon cilantro
6 ounces jalapenos
2 cups white vinegar
2 tablespoons honey
6 teaspoons cumin

Combine all ingredients in large pot. Bring to boil, then simmer uncovered for one or two hours. Place in half-pint or pint sterilized jars, leaving ½ inch headroom. Screw on lids, and boil in hot water bath for 15 minutes. Remove jars from hot water and cool for 12-24 hours, then check lids to make sure they sealed. Makes 13-14 pints.

For a finer product grind these ingredients together.

Dolly Garza shares seaweed with a student at Eielson Elementary School, during Garza's presentation on Native subsistence foods.

Pickled Beach Asparagus I

Large bowl of fresh
  beach asparagus
4 cups white vinegar
4 cups water
2 tablespoons pickling spice

Clean beach asparagus and pack tightly into sterilized pint jars. Bring vinegar and water to a boil. Add spice to water and simmer 5 minutes. Remove pan from heat and pour hot solution over asparagus, leaving ½ inch headroom. Clean rim of jar and screw on lid. Boil submerged in water bath for 20 minutes. Remove and allow to sit for 12-24 hours; then check each jar to ensure it has sealed. Let sit for at least ten days before using to allow the beach asparagus to pickle. Once a jar is opened the unused portion must be refrigerated. Keep jars in cool dry storage for up to a year. Makes about 10 pints.

Pickled Beach Asparagus II

Large bowl of fresh beach
  asparagus, cleaned
  and blanched
6-8 cups white vinegar
4-5 cups white sugar
1 ½ teaspoons turmeric
2 tablespoons mustard seed
1 ½ teaspoons celery seed
½ teaspoon ground cloves
1 large onion, chopped
3-4 red bell peppers, chopped

Clean and soak beach asparagus for 10 minutes in cold water. Drain and pack tightly into sterilized pint jars. Mix vinegar, sugar, and spices in large pot and bring to a boil. Add onion and bell peppers and stir thoroughly. Pour hot solution over beach asparagus, leaving ½ inch headroom. Clean rim of jar and screw on lid. Boil submerged in water bath for 20 minutes. Remove from water and allow to sit for 12-24 hours. Check each jar to ensure it has sealed. Let sit for at least ten days before using to allow the beach asparagus to pickle. Once a jar is opened the unused portion must be refrigerated. Keep jars in cool dry storage for up to a year. Makes about 10 pints.

References

Several of these references are older "classics." You may find them in your library, but probably not in a bookstore. The more recent books such as Druehl or O'Clair and Lindstrom should be available at bookstores and libraries and will prove valuable as you begin to explore seaweeds. Several of the books have excellent recipes that will help pique your interest in harvesting and using local seaweeds.

You may also enjoy another book by Dolly Garza, titled *Surviving on the Foods and Water from Alaska's Southern Shores*. For more field guides and other books and videos about Alaska's seas and coasts, visit the Alaska Sea Grant Web site bookstore at alaskaseagrant.org.

Arasaki, S. and T. Arasaki. 1983. *Vegetables from the Sea*. Japan Publications, Inc., Tokyo, Japan.

Bradford, P. and M. Bradford. 1988. *Cooking with Sea Vegetables*. Healing Arts Press, Rochester, Vermont.

Druehl, L. 2000. *Pacific Seaweeds: A Guide to Common Seaweeds of the West Coast*. Harbour Publishing, Madeira Park, British Columbia.

Ellis, L. 1999. *Seaweed: A Cook's Guide*. Fisher Books, LLC, Tucson, Arizona.

Fortner, H.J. 1978. *The Limu Eater, a Cookbook of Hawaiian Seaweed*. University of Hawaii Sea Grant Program, Honolulu.

References

Lewallen, E. and J. Lewallen. 1996. *Sea Vegetable Gourmet Cookbook and Wildcrafters Guide*. Mendocino Sea Vegetable Company, Philo, California.

McConnaughey, E. 1985. *Sea Vegetables*. Naturegraph, Happy Camp, California.

Mondragon, J.R. and J. Mondragon. 2003. *Seaweeds of the Pacific Coast: Common Marine Algae from Alaska to Baja California*. Sea Challengers, Monterey, California. (ISBN 0-930118-29-4)

National Center for Home Food Preservation, University of Georgia. *www.uga.edu/nchfp*.

O'Clair, R. and S. Lindstrom. 2000. *North Pacific Seaweeds*. Plant Press, Friday Harbor, Washington. (ISBN 0-9664245-1-4)

Scagel, R.F. 1967. *Guide to Common Seaweeds of British Columbia*. British Columbia Provincial Museum, Handbook No. 27.

USDA. 1994. *The USDA Complete Guide to Home Canning*. Publication SF 006, IFAS Publications, University of Florida, P.O. Box 110011, Gainesville, FL 32611, (352) 392-1764.

Waaland, J.R. 1977. *Common Seaweeds of the Pacific Coast*. Pacific Search Press, Seattle. (ISBN 0-914718-19-3)

Index

(*Page numbers in bold indicate illustrations or photos.*)

Alaria marginata 8, **8, 10, 11**
    recipe for 43, 44

beach asparagus (*Salicornia virginica*) 38
    recipe for 49, 50, 53
black seaweed (*Porphyra*) 24, **24, 29, 47, 48, 49**
    recipe for 43, 47, 49
bull kelp (*Nereocystis luetkeana*) 20, **21, 22, 23, 51**
    recipe for 42, 51, 52
bullwhip kelp (*Nereocystis luetkeana*) 20

dulse (*Palmaria mollis*) 30
    recipe for 46

Fucus gardneri 12, **13, 14, 48**
    recipe for 44, 45, 48

herring roe on giant kelp **2**

kelp (*Laminaria*) 16
    recipe for 41, 42
kombu (*Laminaria*) 16
    recipe for 41

Laminaria 16, **17**
    L. bongardiana **17, 19**
    L. saccharina **17, 18**
    L. setchellii **17**
    recipe for 41, 42, 46
laver (*Porphyra*) 24

Index

Native uses of seaweeds 3
Nereocystis luetkeana 20, **21, 22, 23, 51**
    recipe for 42, 51, 52
nori (*Porphyra*) 24
    recipe for 47

Palmaria mollis 30, **30, 31, 33**
    recipe for 42, 46
popweed (*Fucus gardneri*) 12
Porphyra 24, **47, 48, 49**
    recipe for 43, 47, 48, 49
    P. abbottae **24, 25, 26, 27**

Recipes 41-53
References 54
ribbon seaweed (*Palmaria mollis*) 30
    recipe for 42, 46
rockweed (*Fucus gardneri*) 12
    recipe for 44, 45, 48

Salicornia virginica 38, **39**
    recipe for 49, 50, 53
sea lettuce (*Ulva fenestrata*) 34
    recipe for 41
sugar wrack (*Laminaria*) 16

tidal zones, seaweed growth in **1**

Ulva fenestrata 34, **34, 35, 36**
    recipe for 41

wakame (*Alaria marginata*) 8
winged kelp (*Alaria marginata*) 8